Sex Linked Inheritance in Poultry

by T.H. Morgan

with an introduction by Jackson Chambers

This work contains material that was originally published in 1912.

This publication is within the Public Domain.

*This edition is reprinted for educational purposes
and in accordance with all applicable Federal Laws.*

Introduction Copyright 2018 by Jackson Chambers

The World's Largest Selection of Vintage Poultry Books

www.VintagePoultry.com

Self Reliance Books

Get more historic titles on animal and stock breeding, gardening and old fashioned skills by visiting us at:

http://selfreliancebooks.blogspot.com/

Introduction

I am pleased to present yet another title on Poultry.

The work is in the Public Domain and is re-printed here in accordance with Federal Laws.

As with all reprinted books of this age that are intended to perfectly reproduce the original edition, considerable pains and effort had to be undertaken to correct fading and sometimes outright damage to existing proofs of this title. At times, this task is quite monumental, requiring an almost total "rebuilding" of some pages from digital proofs of multiple copies. Despite this, imperfections still sometimes exist in the final proof and may detract from the visual appearance of the text.

I hope you enjoy reading this book as much as I enjoyed making it available to readers again.

Jackson Chambers

SEX-LINKED INHERITANCE IN POULTRY

By T. H. Morgan and H. D. Goodale

(*Presented before the Academy, 8 April, 1912*)

CONTENTS

	Page
Introduction	113
Crosses between Plymouth Rocks and Langshans	114
Description of the breeds	114
History of the breeds	115
Source of breeding stock used	115
Matings	115
Explanation of the symbols used in interpreting the results	116
Parental matings	118
F_1 matings	118
Back matings	120
Summary	121
Description of F_1 Adult Plumage	121
Shank color	123
Booting	125
Down colors	127
Crosses between American Dominique females and Langshan males	128
Parent generation	128
F_1 generation from Langshan ♂ by Dominique ♀	128
F_2 generation	128
Back cross of F_1 ♂ (barred) to Dominique ♀	128
Back cross of F_1, black ♀ to Langshan ♂	129
Other features of the crosses	129
White feathers in wings	130
Color of legs	131
Color of bill	131
Theoretical considerations	131
Bibliography	133

Introduction

In 1908, W. J. Spillman pointed out that, according to a breeder, when Plymouth Rock females are bred to Langshan males all the females are black and all the males are barred. As far as the evidence went, it seemed to show, as he pointed out, that the case was comparable to that of the moth, *Abraxas*, described by Doncaster and Raynor, and of certain crosses among canaries described by Miss Durham.

With the intention of examining further the report cited by Spillman, and of testing, by further combinations, the offspring of the first generation, we began the following experiments in 1909, using Barred Plymouth Rock and Langshan fowls. We undertook also to extend the experiment by using another breed of barred poultry, the American Dominiques.[1] It is currently stated that Dominiques (but not American Dominiques) occur in the ancestry of Plymouth Rocks. We wished to see whether "sex-limited" or "sex-linked" inheritance is found also in this other race. Plymouth Rock-Langshan crosses have been made by one of us (Goodale) on the experimental farm of Mr. B. B. Horton, to whom we are under many obligations for opportunities to carry on the work. The Dominique-Langshan crosses were made by the other (Morgan) at Woods Hole during the summers of 1910-11. In the meanwhile, Pearl and Surface (1910) have described the results of a cross (and its reciprocal) between Barred Rock and Cornish Game. Goodale (1909, 1910) has given briefly some of the results obtained when Barred Rocks are mated (reciprocally) with Buff Rocks and when Brown Leghorns are mated with White Rocks. Hadley (1910) has called attention to similar results published by Cushman in 1893. Davenport (1906, 1909) has described various crosses to one of which certainly (White Cochin by Tosa) and to the others less clearly may be given the same interpretation that applies to the results described in the other papers mentioned above. These crosses all involve the barring factor. Sex-linked inheritance of other factors in poultry has been noted, not only by several of the above writers, but also by Hagedoorn (1909), Sturtevant (1911) and Bateson and Punnett (1908). To Bateson and Punnett is due the explanation of the phenomena of sex-linked inheritance for poultry. More recently (1911) these authors have published a complete account of the inheritance of a factor derived from Brown Leghorns which affects the patency of the type of pigmentation characteristic of the Silky fowl.

CROSSES BETWEEN PLYMOUTH ROCKS AND LANGSHANS

Description of the Breeds.—For a detailed description of the breeds under consideration, reference must be made to the various standard works on poultry. In this paper, only a very brief statement of the chief characteristics involved in the cross will be given.

The Black Langshans (Plate XVII, figs. 2 and 3, and Plate XVIII, fig. 1) are uniformly black, varying somewhat in brilliancy in different regions of the body. The shanks, too, are dull black; the bottoms of the

---

[1] The American Dominique is a younger breed than the Barred Plymouth Rock.

feet are gray. The shanks, moreover, are provided with several rows of feathers, or boot, along the outer edge. The comb is single.

The Barred Plymouth Rocks (Plate XVII, figs. 1 and 4; Plate XVIII, figs. 3, 4, 5, 6) are black and white,[2] the two colors being arranged in alternate bars across the long axis of each feather. The bars vary somewhat in evenness, width and depth of color from individual to individual, and also in different sections of the same bird. Although the American "Standard of Perfection" requires that the two sexes shall be alike in color, the males vary from a darker to a very light color, often appearing very light gray, while the females, though to a less marked extent, vary toward a darker shade. In other words, the breed tends strongly toward a sexual dimorphism of color, with indications of a secondary dimorphism within each sex. The comb is single, and the yellow shanks are free from feathers.

History of the Breeds.—The modern Langshans are the direct descendants of a very old race brought from the interior of China. The Rocks, on the other hand, resulted from a mixture of several races of fowls about forty years ago, which have been gradually brought to a high degree of perfection. The history of the barred character with which we are chiefly concerned is obscure, but evidently it is of very great antiquity, for barred or "cuckoo" birds are to be found in many European countries, Russia included. Brown (1906) states that the plumage of the "Siberian Featherfooted fowl is generally white, whilst others have cuckoo plumage." He notes also that this variety is said to be of ancient lineage. Wright (1902) states that it is probable that the "original Chittagongs, or at least their crossed offspring, were of an 'owl' color as described, probably what we now know as cuckoo or barred." The Chittagongs came from the district of that name in the upper Malay peninsula. An exhaustive search would probably show that barred fowls have been recorded from southeastern Asia.

Source of Breeding Stock Used.—The Langshans came from P. P. Ives of Guilford, Conn. Two of the three Barred Rock males and one of the females were of the well-known Latham strain, but obtained from R. C. Goodale. Four of the barred females were the progeny of the Latham hen by a White Rock male, one was an F_2 from a similar mating, and one was a pure bred female from a Stamford breeder. The White Rock male is known to differ from the barred birds chiefly in the absence of the chromogen factor.

Matings.—In the majority of these matings, the progenies of the individual mothers have not been kept separate. The determinations of the

---

[2] Fanciers prefer to speak of both these colors as grayish.

presence or absence of barring, unless otherwise stated, was made on newly hatched chicks, or those of full term which failed to hatch. This method of determination is made possible by the presence of a gray occipital spot on those chicks which will become barred adults. A full discussion of the point, however, will be given elsewhere. Inroads of vermin, largely rats, have limited the number of which the sex was determined. A description of the adult hybrids is deferred until after all the matings have been described.

Explanation of the Symbols used in Interpreting the Results.—It was pointed out by Spillman, following Bateson, that sex-linked inheritance in poultry could be accounted for on the assumption that the female is heterozygous for sex and the male homozygous, and that when in the female, the barred factor alone is present, it is repulsed by femaleness. We may give this interpretation a more concrete form, if we assume that the factor in question is not carried by the same chromosome that carries the factor for the female sex; *i. e.*, in the heterozygous female the chromosome that carries femaleness also lacks the factor for barring, and its mate that lacks the factor for femaleness carries the factor for barring. No interchange between the chromosomes (if two really exist) can take place, perhaps because they fail to pass through those stages in synezesis when such a process becomes possible.

If F = female, f its absence or male; B = barred, b its absence; N = black, then the formulas for the barred fowls will be:

Barred	♀	FNb	fNB
"	♂	fNB	fNB

For the Langshan fowls, the formulas will be:

Black	♀	FNb	fNb
"	♂	fNb	fNb

Whether the female-producing gamete of the Barred Rock really carries black or only the absence of barring will not affect the nominal results here recorded, but other experiments to be described by Goodale will show that "black" is probably present.

In order to see how these formulas apply to the crosses under consideration, let us take first the case of the cross between the Langshan hen and the Plymouth Rock cock (fig. 1). The formulæ are as follows:

	Langshan ♀		FNb	fNb
	Barred Rock ♂		fNB	fNB
F_1	Barred ♀		FNb	fNB
	" ♂		fNb	fNB

F₂	Black ♀	FNb	fNb
	Barred ♀	FNb	fNB
	" ♂	fNB	fNB
	" ♂	fNB	fNb

Barring is dominant to self color, as is shown in the last case, where in the F₁ generation all the offspring are barred. In the second generation,

FIG. 1.—*Cross of Plymouth Rock ♂ to Langshan ♀*

there occur barred ♂ and both barred and black ♀. The grandmaternal color, black, appears in the grand-daughters and not in the grandsons.

The reciprocal cross between the Barred Rock female and the Langshan male (fig. 2) may be represented as follows:

	Barred Rock ♀	FNb	fNB
	Langshan ♂	fNb	fNb
F₁	Black ♀	FNb	fNb
	Barred ♂	fNB	fNb
F₂	Barred ♀	FNb	fNB
	Black ♀	FNb	fNb
	Barred ♂	fNb	fNB
	Black ♂	fNb	fNb

Parental Matings.—(1) From the five Langshan hens by a Barred Rock male, there were 34 young, all barred: 12 were females and 8 were males (fig. 1):

FNb	fNb	....	Langshan ♀
fNB	fNB	....	Barred Rock ♂
FNb	fNB	....	Barred ♀ 12
fNB	fNB	....	" ♂ 8

Fig. 2.—*Cross of Langshan ♂ to Plymouth Rock ♀*

(2) From the various barred females bred to a Langshan male, there were 20 barred (15 ♂) offspring and 25 black (16 ♀) (fig. 2):

FNb	fNB	....	Barred ♀
fNb	fNb	....	Langshan ♂
FNb	fNb	....	Black ♀ 16
fNB	fNb	....	Barred ♂ 15

F₁ Matings.—(3) Four barred F₁ females from (1) were bred to a barred male, No. 568, from (2). This was done because the only adult male, No. 784, from (1) did not mature until long after his sisters were laying, while a change of residence on the part of the writer prevented the accomplishment of the *inter se* mating. From the cross-mating, however, there were 25 barred (12 ♂ and 9 ♀) offspring and 13 black (8 ♀) (fig. 1). Expectation on the Spillman-Bateson hypothesis is 28½-9½. One individual, a male, with the gray spot reduced to a few plumules was excluded from the count as doubtful.

$$F_1 \begin{cases} FNb \quad fNB \dots\dots\dots\dots\dots\dots \text{Barred ♀} \\ fNB \quad fNb \dots\dots\dots\dots\dots\dots \text{`` ♂} \end{cases}$$

$$F_2 \begin{cases} FNb \quad fNB \dots\dots\dots\dots\dots\dots \text{Barred ♀} \quad 9 \\ FNb \quad fNb \dots\dots\dots\dots\dots\dots \text{Black ♀} \quad 8 \\ fNB \quad fNB \dots\dots\dots\dots\dots\dots \text{Barred ♂} \\ fNB \quad fNb \dots\dots\dots\dots\dots\dots \text{`` ♂} \end{cases} \Big\} 12$$

FIG. 3.—*Cross of barred ♂ to F_1 black ♀*

(4a) The 6 black females from (2) were mated to a litter barred brother, No. 569, giving 41 young; 16 of which were black (8 ♂, 6 ♀) and 25 barred (6 ♂, 10 ♀). (4b) Later, they were bred to No. 784, giving 22 young; 7 were black (0 ♂, 1 ♀) and 15 barred (5 ♂, 1 ♀). The combined results of these matings were 63 young, of which 23 were black (8 ♂, 7 ♀) and 40 barred (11 ♂, 11 ♀) (fig. 2). The departure from the expected ratio of 31½ is considerable.

$$
\begin{array}{lllll}
F_1 & \{ & FNb & fNb & \ldots\ldots\ldots\ldots\ldots\ldots \text{Black } \female \\
    &   & fNB & fNb & \ldots\ldots\ldots\ldots\ldots\ldots \text{Barred } \male \\
\hline
    & \{ & FNb & fNB & \ldots\ldots\ldots\ldots\ldots\ldots \text{Barred } \female & 10 + 1 \\
F_2 &   & FNb & fNb & \ldots\ldots\ldots\ldots\ldots\ldots \text{Black } \female & 6 + 1 \\
    &   & fNb & fNB & \ldots\ldots\ldots\ldots\ldots\ldots \text{Barred } \male & 6 + 5 \\
    &   & fNb & fNb & \ldots\ldots\ldots\ldots\ldots\ldots \text{Black } \male & 8 + 6 \\
\end{array}
$$

Back Matings.—(5) The 6 black F_1 females used in (3) and (4) were bred to a pure Barred Rock male and gave 9 barred young, fulfilling expectation (fig. 3).

$$
\begin{array}{llll}
FNb & fNb & \ldots\ldots\ldots\ldots\ldots\ldots \text{Black } \female \\
fNB & fNB & \ldots\ldots\ldots\ldots\ldots\ldots \text{Barred } \male \\
\hline
FNb & fNB & \ldots\ldots\ldots\ldots\ldots\ldots \text{Barred } \female \\
fNb & fNB & \ldots\ldots\ldots\ldots\ldots\ldots\ \text{"} \ \male \\
\end{array}
$$

Fig. 4.—*Cross of Langshan $\male$ to F_1 barred $\female$*

(6) The 4 barred F_1 females used in (3) were bred to a pure Langshan male. Of the 30 offspring, 14 were barred (1 $\male$, 1 $\female$) and 16 black (0 $\male$, 3 $\female$). This corresponds closely to the expected equality in ratio (fig. 4), except for the possible barred female. The determination of sex in this case was made on rather poorly preserved material.

FNb	fNB	.... Barred ♀
fNb	fNb	.... Black ♂
FNb	fNb	.... Black ♀
fNB	fNb	.... Barred ♂

(7) Several of the parental barred females were mated with No. 568, barred F_1 ♂, giving 8 barred (3 ♀, 1 ♂) and 3 black (1 ♀).

FNb	fNB	.... Barred ♀	
fNB	fNb	.... " ♂	
FNb	fNB	.... Barred ♀	3
FNb	fNb	.... Black ♀	1
fNB	fNB	.... Barred ♂	
fNB	fNb	.... " ♂	} 1

(8) Four of the parental Langshan females mated to No. 569, barred F_1 male, gave 17 barred (6 ♀, 2 ♂) and 13 black (4 ♂; no record for ♀).

FNb	fNb	.... Black ♀	
fNB	fNb	.... Barred ♂	
FNb	fNB	.... Barred ♀	6
FNb	fNb	.... Black ♀	?
fNb	fNB	.... Barred ♂	2
fNb	fNb	.... Black ♂	4

(9) Two of the barred F_1 females used in (3) were bred to a Rock male, not, however, of the Latham strain. There were only 5 chicks, all barred.

FNb	fNB	.... Barred ♀
fNB	fNB	.... " ♂
FNb	fNB	.... Barred ♀
fNB	fNB	.... " ♂

Summary.—Expectation in all these matings has been closely fulfilled (with the exception of No. 4 and perhaps No. 6), on the assumption that the barred female is heterozygous for both barring and sex, both femaleness and barring being dominants and that the two factors do not occur in the same gamete.

Description of F_1 Adult Plumage.—The males (Plate XVII, figs. 5 and 7) resemble one another closely and together with the barred F_1 females are very suggestive of the Coucou de Malines, a Belgian breed. The barring of the individual feathers of the males (Plate XVIII, figs. 10, 13) is less sharp and regular than that of the parental Rocks (Plate XVIII, fig. 8), while the dark bars tend to run together, particularly in wings and tail, and at the same time, the light bars become more or less smoky. The primaries, indeed, can be called barred only by courtesy, for the light bars are only represented by white splashes along the shaft (Plate XVIII, fig. 14). This region, however, is one in which the fanciers have found great difficulty in producing even and regular barring.

One barred male is particularly interesting in that a few feathers show distinctly the Jungle coloration (Plate XVIII, fig. 12) which probably exists as a cryptomere in the Langshans.

All the males have numerous feathers wholly or partly black (Plate XIX, figs. p-u) and this is true also for the barred females. The last, except for the black feathers, are well barred (Plate XVII, fig. 6) and can scarcely be distinguished from the parental stock. Even the individual feathers, except the remiges and retrices in which the bars run together, conform closely to the pattern shown by many thoroughbred Barred Rocks.

The color of the F_1 black females (Plate XVII, fig. 8) is indistinguishable from the parental Langshans.

Comparatively few members of the F_2 generation reached maturity. The only points of particular interest are the appearance of very light as well as dark males, of both black and barred males having a few feathers showing the Jungle fowl coloration and of dark-colored females with the barring somewhat blurred.

The non-appearance of game (Jungle) colored birds in F_2 is due presumably to the fact that black is duplex in both Rocks and Langshans, and thus the Jungle fowl color is concealed. There are, however, indications that black may sometimes exist in a simplex condition among Rocks; so that, if suitable matings were made, the Jungle fowl color might appear. Unless the occasional feather showing Jungle fowl color is due to a simplex condition of black, its appearance may mean that hybridization in some way has upset the usual complete dominance of duplex black over the Jungle fowl coloration.[3]

As already stated, Pearl and Surface have published their results in crossing Plymouth Rocks and Cornish Indian Game. Our results entirely accord with theirs, as far as inheritance of barring is involved. They classify their birds as barred and non-barred, ignoring intentionally the differences among the non-barred birds. Our results are simpler, in so far as all our non-barred birds are black, but the principle involved is the same in both cases. Pearl and Surface have also made all possible back crosses between the parents and the F_1 generation. Our results are in entire harmony with theirs, but they have the advantage of a larger number of offspring in their matings.

---

[3] DAVENPORT, 1909, p. 72.

Shank Color.—The color of the shanks of all chicks hatched was recorded, but the color often changes as the birds become older, so these records prove to be of small value. The change in shank color is particularly characteristic for the class called yellowish or flesh-colored black. This class may give rise to all three of the adult shank colors recognized, black, gray and yellow. Black-shanked chicks seem always to develop into black-shanked adults, and while yellow-shanked chicks probably do not produce black-shanked adults, they may give rise to either yellow or gray-shanked adults.

The infantile shanks among the F_2 not only show the expected classes, but these classes pass by imperceptible grades into one another. Frequently, one part of the shank, particularly the toes, differs from the remainder; while, in many cases, the distribution of color forms a mottled pattern. The distribution of color upon the toes is likewise extremely variable and often asymmetrical. In almost every case, however, some part of the toes is flesh or yellow. This variation is due, presumably, to some extension or restriction factor. Similar variation in the distribution of color in F_1 was also recorded.

The shank color of the F_1 adults falls into two classes, black and gray. The term gray is used rather loosely to cover a particular though somewhat variable coloration of the shanks. At a distance, the shanks do indeed appear gray just as a Barred Rock appears gray, and just as the "gray" of the Rocks resolves into a pattern on closer inspection, so the gray of the shanks is not a single or uniform color. For convenience of description, we may say that the ground color is steel gray, variously mottled with patches of darker gray or of black. Parts of the shank often have a bluish cast. The posterior side and particularly the bottoms of the feet are somewhat flesh colored. Mottling does not as a rule occur on the bottom of the feet, so that though the term gray is applied to them in a later paragraph, it is to be understood that they do not have the same appearance on the shanks proper but rather are a grayish flesh, self color. The three classes of black, gray and yellow do not grade into each other.

The six F_1 black females had black shanks. The three males and the four barred females had gray shanks. Apparently, we have here a case of sex-linked inheritance. This, however, may not be the case but may be due rather to the black spreading over onto the shanks just as it often spreads over onto the comb. In the barred birds, we may suppose that the barring factor operates to prevent the spreading of black over the shanks, just as it also produces the characteristic barring of the feathers of the boot. Thus, the colors hypostatic to black are revealed. However this may be, in F_2 the black birds again have black shanks, but the bottoms of their feet, which are usually incompletely covered by black, are either gray or yellow. The allelomorphs involved, then, are gray versus yellow (or $Gray_2$, $Yellow_2$, X, No $gray_2$, $Yellow_2$), the latter being recessive to the former. Moreover, among the barred birds, only gray or yellow shanks appear, or in other words, gray-shanked birds always have gray soles, yellow shanked birds yellow soles, but black-shanked birds may have either gray or yellow soles.

Since, then, the black-shanked condition is due to an extension of the general black color of the body, we need consider further only the relation of gray to yellow, the determinations being made, of course, only on the bottoms of the feet and when the birds were several months old. In F_1, there were only gray or pinkish gray feet, and, therefore, there is no evidence that gray is sex-linked. Moreover, since no other color than yellow appeared in F_2, yellow is probably common to both Langshan and Rocks, so that absence of gray in this case means yellow. In F_2, not all the adults were available for study, as the importance of foot color was not realized until after many of the birds had been disposed of, but in 17 cases, 13 were gray and 4 yellow. The back mating of F_1 gray male to P_1 gray (Langshan) female gave 6 gray. The back mating of P_1 yellow (Rock) male to F_1 gray (black plumage) female gave 6 gray to 2 yellow. These results indicate, then, that gray and yellow feet (or shanks, leaving out of consideration the supermelanic coat) behave in simple Mendelian fashion.

We have suggested that black individuals have black shanks, because a restriction factor is absent from these birds, so that the body color spreads out as a self color over the shanks. Such a "restriction" factor would be sex-linked. Is it, then, the same as the barring factor? If it were a separate factor, we should expect that, in F_2, a certain amount of segregation would take place. This has not been observed, so that it seems probable that the black shanks of the black birds are due to the absence of the barring factor and the mottled shanks to its presence, unless some "association" exists. Thus, the presence of the barring factor results in two (perhaps three) distinct somatic conditions, viz.: barred feathers and mottled shanks, and, as a possible third, the gray occipital spot of young chicks. In other words, we have two or more unit characters resulting from the operation of a single factor.

There are some considerations of a practical nature resulting from the relations between shank color and sole color which should be mentioned. If the black color covered the entire foot, we should be unable to determine what color underlay the black, except perhaps by long-continued breeding tests. Gray would, therefore, appear to be a sex-linked character. In F_2, however, the results would appear peculiar, for while we should have the three classes of black, gray and yellow shanks, the black shanks would always appear associated with black birds, while gray and yellow shanks would go with barred birds. This conclusion does not agree with the results expected when two independent sex-linked characters are involved. In F_3, the observed results would be very complicated. A discussion of the various possible explanations which might be devised to meet the situation would hardly be profitable here, but a comparison of the results expected when the color of the soles of the feet is taken into account with those when they are omitted may furnish the key to similar cases.

Booting.—The Barred Rocks are typically clean shanked, but occasionally a bird is found with a few "stubs." The boot of the Langshan corresponds approximately to that shown in many of the older pictures of Cochins and Brahmas and may perhaps be regarded as the primitive type from which the modern highly developed boot of Cochins has been developed.

For the F_1 generation, booting was recorded on the chicks as "present" in all cases but two. These two occur among the first four recorded, so that it is possible that, if only a few stubs were present, they may have been regarded as slightly atypical clean shanks. In one other case, booting was nearly absent. Of the 13 adults, the three males and four barred females were alike in that the amount of booting was decidedly scanty, being reduced to about two or three imperfect rows of rather short feathers. The six black females were more variable, due apparently to greater variation in length of feather rather than to variation in the number of rows, the result being a greater variation in amount of boot.

A much larger range in the amount of booting appeared in the next generation. The following relative grades of boot in the chicks were recognized: A, B, C, D, E and absent. No emphasis is to be laid on these degrees, except in so far as they show the general distribution of boot. A and B correspond approximately to that of the parental Langshan, and C and D to that of the F_1 hybrids. Among the adults, not only were there some birds heavily booted like the Langshans, some like the hybrid and others clean-shanked like the Rocks, but one bird had two rows of rather long feathers and one bird four rows of short feathers, indicating that there is more than one component to boot.

TABLE I
Distribution of Booting in F_2 and $F_{1.5}$

Mating No.	A	B	C	D	E	Absent	Total	Remarks
3	1	7	1	14	0	6	29	F_1 females from 1 × F_1 male from 2.
4a	0	10	2	8	3	4	27	F_1 females from 2 × litter brother.
4b	0	6	3	4	5	3	21	F_1 females from 2 × reciprocal litter brother.
5	0	0	0	0	2	6	8	F_1 females from 2 × Rock male.
6	0	7	10	11	2	0	30	F_1 females from 1 × Langshan male.
7	0	0	0	3	4	3	10	Female Rocks × male from 1.
8	3	13	4	10	0	0	30	Female Langshans × male from 1.

TABLE II

Mating No.	Generation	Expectation			Observed		Total	Remarks
		Per cent clean	Based on total recorded					
			Clean	Booted	Clean	Booted		
1	F_1	0	0	all	2*	24	26	Langshan females × Rock male.
2	F_1	0	0	all	0	32	32	Rock females × Langshan male.
3	F_2	18.75	5.4	23.6	6	23	29	Females from 1 × male from 2.
4a & b	F_2	12.5	6.	42.	7	41	48	Females from 2 × males from both 1 and 2.
5	$F_{1.5}$	50	4.	4.	6	2	8	Females from 2 × Rock male.
6	$F_{1.5}$	0	0	all	0	30	30	Females from 1 × Langshan male.
7	$F_{1.5}$	37.5	3.75	6.25	3	7	10	Female Rocks, male from 1.
8	$F_{1.5}$	0	0	all	0	30	30	Female Langshans, male from 1.

The results are in entire agreement with Davenport's and confirm his theory of an inhibitor. The back matings suggest that the amount of boot varies with the increase or decrease in the amount of booted "blood". There are, however, one or two other theoretical ways of accounting for the observed facts. If we assume that booting is common to both Langshans and Rocks and is recessive to a pair of complementary factors, both

* See above in text.

of which must be present and one of which must be duplex in order to bring about a complete suppression of the booting, the outcome approximates the observed ratios of booted to non-booted.

By assuming that the factor which exerts its effect in either the duplex or simplex condition is sex-linked, the results shown in Table II are obtained. The distribution of the sexes is not given, because the numbers available are inadequate for the solution of a problem as complex as the present one. While the correspondence between theory and observation in this case is close, an attempt to apply it to Davenport's data resulted in only partial success. This may mean only that more or different factors are involved in the production of boot in Brahmas, Cochins and Silkies than in Langshans, or that the factors causing the inhibition of boot development in Plymouth Rocks are different from those of Tosa, Minorca and other smooth-shanked birds used by Davenport. Among possible factors concerned in boot production should be included those general factors which affect feather growth, in the same way as barring or other color factors control the color of the feathers of the boot as well as those of the body.

Down Colors.—The Langshan chick is black dorsally but yellowish white beneath and has white wing tips. The white ventral area often extends upwards, particularly on the head, so that in some cases in this region only the crown and nape remain black. The white area of the wing tips at the same time increases in size, so that the black dorsal surface becomes reduced in amount.

The Rock chick, however, though black dorsally except for the gray occipital spot, is usually dark gray beneath, but very often there are several light gray or white areas, which occasionally become more or less confluent, and in extreme cases most of the ventral surface is white and to a limited degree overlaps the Langshan type.

In classifying the chicks, all were called "black," i. e., of Langshan type, in which at most the breast region was partly pigmented. This region in the Barred Rock chick is the last to lose pigment. All others were classified as "barred". While this mode of treatment proved to be inadequate for the entire solution of the inter-relations of these characters, it was found, first, that both types appear in F_1, but that the "blacks" are far more numerous than the "barreds"; second, that "blacks" F_1 interbred or backmated throw some "barreds", but not in simple Mendelian proportions.

CROSSES BETWEEN AMERICAN DOMINIQUE FEMALES AND LANGSHAN MALES

Parent Generation.—Both the hens and the cock were purchased from breeders of these strains.[5] The one peculiarity calling for notice is the occasional occurrence in the Dominique hens of black or partly black feathers (Plate XIX, figs. *b, d, e*). One of the four hens used had several such feathers. The other hens were free from them. The American Dominiques have barred feathers (Plate XIX, figs. *a, c*), essentially like those of Plymouth Rocks.

F_1 *Generation from Langshan ♂ by Dominique ♀.*—About 15 offspring were reared; the hens were black and the cockerels barred. Of these, five hens and two cocks were bred from. The black hens were like the father as to color; the males were barred like the mother, except that a large number of black feathers were present—some feathers entirely black (Plate XIX, figs. *r* and *t*) and others barred and black (Plate XIX, figs. *p, q, s, u*).

F_2 *Generation.*—In the second generation, there were recorded 15 blacks and 14 barred birds. Three of the latter died or were killed by animals. Of the remaining, there were 11 male and 15 females tabulated as to color as follows:

	♀	♂
Barred	8	4
Black	7	7

The barred birds were fairly uniform. They were kept for about two months, when their feathers were well developed. A few birds were distinctly darker than the rest, and one bird was much lighter. Certain details regarding white feathers in the wings will be spoken of later.

Back Cross of F_1 ♂ (Barred) to Dominique ♀.—One of the sons was crossed to the four hens that had produced his generation. A first census of the offspring, when the birds were small, gave 19 barred and 4 black birds. A later count when the birds were older gave 14 barred and 4 black. Five barred birds had disappeared. The distribution of color and sex of 16 of these birds was as follows:

	♀	♂
Barred	7	5
Black	4	0

---

[5] The Dominiques came from W. H. Davenport, Colrain, Mass. The source of the Langshans is given on page 115.

There were also two barred birds whose sex was omitted by mistake in the records. The expectation is three barred to one black, which is closely realized.

Back Cross of F_1 Black ♀ to Langshan ♂.—Five black hens were bred to a Langshan cock of the same strain but not the actual father of these hens. Another black hen that came from a similar cross with a Plymouth Rock was also present in the same pen, so that some of the offspring may have come from this hen also. There were 18 black young, of which 11 were males and 7 females. In addition, however, there was one barred chick. Now, the black hens had been with a barred cock to give the F_2 generation. They had been for three weeks with the Langshan male before the eggs fertilized by the black cock were kept for incubation. There can be little doubt that one of the spermatozoa of the original male had carried over and produced this bird. If this case is thrown out, the results are consistent.

Other Features of the Crosses.—The Langshans have feathered tarsus (booted); the Dominiques have clean shanks. All of the F_1s recorded were booted, though not strongly. In the F_2 generation, there were 14 booted and 11 clean shank, distributed as follows:

	Booted		Clean	
	♀	♂	♀	♂
Barred	7	2	1	2
Black	1	4	5	3

It is clear that booted shanks dominate[6] but imperfectly in this cross, as in other crosses of poultry. Some of the F_2 offspring had heavily booted legs; others were like the F_1 generation. No sharp line between the classes in the F_2 generation could be drawn.

There is no evidence of any association here between black and booted (the paternal combination) and barred and clean-shanked (the maternal combination).

When the barred and booted F_1 male was bred to four Dominique hens, the results are shown in the next table:

	Booted		Clean	
	♀	♂	♀	♂
Barred	6	1	1	4
Black	1	0	3	0

---

[6] In the sense that an inhibitor is present in clean shanks.

In this case, the male was heterozygous for condition of tarsus; the hens pure and recessive. The result calls for equal distribution of booted and clean shanks unless "association" occurs. The numbers are too small to have any significance. Even as they stand, however, they have no meaning, if coupling be made responsible for the distribution of the characters.

When the Langshan male was crossed to the black hens (both sexes booted, but the hens heterozygous) all of the offspring were booted, which is in accordance with dominance of booted shanks.

White Feathers in Wings.—In the F_2 young birds, the presence of white and partly white feathers in the wings was noticed (Plate XIX, figs. *f-n, o, v* and *w*). They were most obvious in the black birds, perhaps because of the sharper contrast. These feathers are some of the primaries and a few of the coverts at the base of the primaries. As shown in Plate XIX, figs. *f* to *k*, they are rarely pure white, but often mottled or splotched. They were not recorded in the F_1 birds, and if present they were overlooked. The records of birds without and with these white feathers were as follows:

	♀	♂
Barred, no white	6	4
" with "	2	0
Black, no white	6	4
" with "	1	3

In all, there were 20 chicks without and 6 with white feathers. This looks like a case of Mendelian inheritance, but it may be purely a coincidence. We do not know how often such feathers occur in chicks of the original breeds, or whether they are only juvenile, or physiological effects of the condition of the bird. Probably they would have disappeared in later molts, had the chicks been kept longer.

When the Langshan cock was bred to the black F_1 hens, 4 of the chicks had no white and 14 had white in the wings. If the black male is heterozygous for this condition, the result is not in accordance with the assumption that this is a Mendelian recessive character.

When the Dominique hens were bred to the F_1 barred males, there was no white in the 15 recorded offspring. This result is not in harmony with the same supposition, but the black male used in the last experiment was not the same father as for the barred males of the first cross. The father of the barred male in the first case was a brother of this one. It is still possible, therefore, that one male was homozygous and the other heterozygous for the white-feathered condition. Without, however, fuller information, not much weight can be given to these results.

Color of Legs.—It has been stated by Bateson and by Pearl that yellow and black shanks in certain breeds of poultry show "sex-linked" inheritance. This is not apparent in the Langshan-Dominique crosses, except in so far as black shanks accompany black color of feathers. For example, in the F_2 generation, there are recorded 13 black birds with black legs. Of these, 5 were deep yellow on the under side of the feet. In addition, there was one male that had yellow shanks and yellow under the feet. There were recorded 12 barred F_2 chicks with yellow shanks. Of these 12 birds, 4 are recorded as having very pale yellow or whitish legs. It would appear from this case that black and yellow shanks accompany black and barred plumage, at least as a rule.

In the back cross of the F_1 barred male to the parent Dominique hens, in which there were barred males and females and only black females, all 4 of the black birds had black legs, while all 12 barred birds had yellow or pale legs. Among these 12 barred birds, there were 5 with pale legs; in 2 of these and in one yellow, there were spots of black or dark color, at least on the tarsus.

These rather meager figures, as far as they go, show that shank character and color of plumage go together, and that black shanks and yellow shanks are only an accompaniment to sex-linked inheritance of plumage. The data are manifestly few, however, and it may well happen that the two characters may appear disassociated.

Color of Bill.—The color of the bill seems to run a parallel course. Full records for the back cross given above were kept. Here, 13 barred birds had yellow bills and 5 black hens had black bills, but one of the latter had much yellow on it, and two of the former had black: one was black with yellow tip and the other was yellow and black. There is much variability in the color of the bill, and the above statements are insufficient to warrant any generalizations.

THEORETICAL CONSIDERATIONS

The current formula for sex inheritance in fowls represents the female as heterozygous for sex, F-O, and the male homozygous, O-O. If F is identified with a special chromosome connected with sex determination, the formula calls for one more chromosome in the female than in the male. At present, evidence on this point is conflicting and insufficient. It is true that Guyer has described two kinds of spermatozoa in the male, one with an X and one without. If this X is the same as in other animals, then the spermatozoa containing it must be female producing, and the female should contain one more chromosome than the male. This means that the male and not the female is heterozygous for sex. The experimental evidence is flatly opposed to this latter interpretation, and, therefore, until Guyer's evidence is confirmed or refuted, the case must be left open.

On the other hand, if, as the experimental evidence shows, barring is "repulsed" by femaleness and if both of these factors are carried by chromosomes, the formulas are deficient in having no chromosome to carry barring,—a contradiction of terms. It may be, however, that the X-chromosome in fowls has a mate which we may call Y which would carry barring but not femaleness. The formulæ would then be:

$$\text{Female} \dots \dots X - Y$$
$$\text{Male} \dots \dots Y - Y$$

$$\begin{array}{ll} XY & \text{Female} \\ YY & \text{Male.} \end{array}$$

On this interpretation, the factor for femaleness would be contained in X but absent from Y, while barring is contained in Y. This scheme is compatible with the experimental evidence and gives consistent results for all combinations.

The irregularities that have been observed in the "reduction division" both in birds and in man suggest the possibility that the sex chromosomes are united to other chromosomes as in some other animals. If the union is variable, as in the nematodes, it may be that the X and the Y (if Y exists) may sometimes pass to the poles of the spindle during reduction in conjunction with other chromosomes and sometimes be free to pass to the poles independently. If further study should establish this view, it will have a very direct bearing on the relations discussed above. If the factor F for femaleness is carried by chromosomes attached to one member of another pair, the mate of this member may be the chromosome that carries the factor for barring. If this were the case, however, interchange between these two members would lead to the barring factor being transferred to the chromosome attached to the sex chromosome. This is in contradiction to the experimental evidence which would lead rather to the conclusion that a Y element lacking the factor for barring is present. The Y may be attached to the mate of the chromosome carrying the sex factor.

At present, only a few cases have been discovered in which a sex-linked character is dominant, viz.: in fowls and in one character in *Drosophila*. The only other cases, besides the one in poultry in which sex-linked inheritance occurs and sex is heterozygous in the females, is that of *Abraxas* and that of canaries. In both of the latter, the sex-linked factor is recessive. There are no *a priori* grounds why a character of this sort may not be dominant, if some other Mendelian characters may also be dominant.

The factor for black, N, is treated in our formulæ as present in all of the gametes both of the female and of the male. It is not allelomorphic to barring, B, although its presence in the female-producing egg when barring is present in the correlated male-producing egg may appear to bear this interpretation. From the chromosome point of view black may be, so far as we know, in other chromosomes than those carrying barring; hence its more general distribution.

BIBLIOGRAPHY

American Poultry Association, The American Standard of Perfection. 1905.

BATESON, W.: Facts Limiting the Theory of Heredity. Science, N. S., Vol. XXVI. 1907.

——: Mendel's Principles of Heredity. 1909.

——: The Inheritance of the Peculiar Pigmentation of the Silky Fowl. Jour. Genet., Vol. I. 1911.

BATESON, W., and PUNNETT, R. C.: The Heredity of Sex. Science, N. S., Vol. XXVII. 1908.

BROWN, E.: Races of Domestic Poultry. London. 1906.

DAVENPORT, C. B.: Inheritance in Poultry. Carnegie Pub. No. 52. 1906.

——: Inheritance of Characteristics in Domestic Fowl. Carnegie Pub. No. 121. 1909.

GOODALE, H. D.: Sex and Its Relation to the Barring Factor in Poultry. Science, N. S., Vol. XXIX. 1909.

——: Breeding Experiments in Poultry. Proc. Soc. Exp. Biol. and Med. Vol. VII. 1910.

HADLEY, P. B.: Sex-limited Inheritance. Science, N. S., Vol. XXXII. 1910.

HAGEDOORN, A. L.: Mendelian Inheritance of Sex. Archiv. Ent. Org., Bd. XXVIII. 1909.

PEARL, R., and SURFACE, F. M.: Studies on Hybrid Poultry. An. Rep. Maine Agric. Exp. Station. 1910.

——: On the Inheritance of the Barred Color Pattern in Poultry. Archiv. Ent. Org., Bd. XXX. 1910.

——: Further Data Regarding the Sex-limited Inheritance of the Barred Color Pattern in Poultry. Science, N. S., Vol. XXXII. 1910.

SPILLMAN, W. J.: Spurious Allelomorphism; Results of Some Recent Investigations. Am. Nat., Vol. XLII. 1908.

STURTEVANT, A. H.: Another Sex-limited Character in Fowls. Science, N. S., Vol. XXXIII. 1911.

TEGETMEIER, W. B.: The Poultry Book, etc. London. 1867.

WRIGHT, L.: The New Book of Poultry. London. 1902.

PLATE XVII

BARRED PLYMOUTH ROCKS, LANGSHANS AND THEIR CROSS-BRED OFFSPRING

Fig. 1. Barred Plymouth Rock cock.
Fig. 2. One of the parental Langshan hens.
Fig. 3. Langshan cock. Stock of Mr. Ives.
Fig. 4. One of the parental barred hens. This particular hen is an F_1 from White Rock male X Barred Rock female.
Figs. 5 and 6. The F_1 from Barred Rock cock by Langshan hen.
Figs. 7 and 8. The F_1 from Langshan cock by Barred Rock hen.

ANNALS N. Y. ACAD. SCI.

VOLUME XXII, PLATE XVII

PLATE XVIII

FEATHERS OF LANGSHANS, BARRED PLYMOUTH ROCKS AND THEIR OFFSPRING

Hackle feathers (except 11 and 14) from the various types of birds used in the experiments.

FIG. 1. From Langshan female.
FIG. 2. From F_1 black females.
FIGS. 3 and 4. From a pure bred Barred Rock female.
FIG. 5. From a second Barred Rock female. Note the differences in the evenness of the barring.
FIG. 6. From the hen shown in Plate XVII, Fig. 4.
FIG. 7. From F_1 barred female.
FIG. 8. From a pure bred Barred Rock cock.
FIG. 9. From same bird, illustrating a partially black feather occurring in pure bred stock.
FIGS. 10, 11, 12, 14. From F_1 bird shown in Plate XVII, Fig. 5.
FIG. 10. Hackle feather.
FIG. 11. Breast feather.
FIG. 12. Shows the Jungle fowl coloration.
FIG. 13. From bird shown in Plate XVII, Fig. 7.
FIG. 14. Primary, to show reduction in barring.

PLATE XIX

FEATHERS OF AMERICAN DOMINIQUE FOWL

FIG. *a*. Dominique hen. Barred feather.
FIG. *b*. Dominique hen. Black feather.
FIG. *c*. Dominique hen. Barred feather.
FIG. *d*. Dominique hen. Black feather.
FIG. *e*. Dominique hen. Black feather.
FIGS. *f-k*. Consecutive primaries of barred F_2, showing four nearly white feathers.
FIGS. *l-n*. Three white covert feathers from wing.
FIG. *o*. Consecutive primaries of black F_2, showing some white at tip.
FIGS. *p-u*. Tail coverts of F_1 barred male, showing uniform black regions and barring in same feather.
FIG. *v*. Wing of black F_2, showing white tips to primaries and two white covert feathers at their base.
FIG. *w*. Wing of barred F_2, showing a nearly white primary and irregular barring on some other feathers.

www.ingramcontent.com/pod-product-compliance
Lightning Source LLC
Chambersburg PA
CBHW062345220526
45469CB00008B/2841